Astromedicine: redefining Space Medicine

ASTROMEDICINE
REDEFINING SPACE MEDICINE

Vanessa Farsadaki

BSc, MS, MSc, MBA, EMGM, CBT, MD

Astromedicine: redefining Space Medicine

Copyright © 2023 by Vanessa Farsadaki

Cover copyright © 2023 by Vanessa Farsadaki

Astromedicine is trademarked under Space Exploration Strategies

All rights are reserved.

V. Farsadaki

To all those who dared to open a new path

ACKNOWLEDGMENTS

Thank you to all the people who inspired me growing up, from teachers to professors and from family members to individuals I met once but was sufficient for them to give me a single piece of information that shaped this book. Above all, thank you to those that doubted me and hoped they'd bring me down. Thank you to my Greek heritage that taught me how to dare and my American heritage that taught me how to serve. Thank you to you, reader who entrusted me with your time to start a revolution.

V. Farsadaki

INTRODUCTION

Why? Have you ever stopped for just a moment to consider this remarkably elegant, extremely powerful, extremely compelling three letter word? In posing the question "Why" to ourselves, we can open doors to a universe that can tell us all of its wonders. It can drive us to seek ultimate truths; it can propel us to loft heights. It can lead to solutions to the many ills that plague us. It was the word "Why" that drove me to write this book. The question that I assigned to that beautiful word was this:

"With humanity rapidly preparing to take its place amongst the stars, why have we not recalibrated our medical thinking to seriously address the challenges that will kill us out there if we are not prepared."

That's what this book is all about: Addressing the need to change the conversation about how medicine should evolve to support

Astromedicine: redefining Space Medicine

humanity's future space faring needs.

Consider the following historical context. "For ages, humanity has been driven to explore the unknown, find new worlds, test the limits of science and technology, and then push even farther. Our culture has benefited much from the intangible drive to explore and question boundaries.

Human space exploration contributes to the understanding of fundamental issues about our location in the universe and the history of our solar system. By addressing the obstacles of human space travel, we may advance technology, establish new industries, and contribute to the peaceful coexistence of nations.

The program represents a breakthrough in the way NASA explores space in 1992. The goal was to learn more about the solar system by examining the planets, their moons, and minor celestial bodies like comets and asteroids.

Discovery missions have accomplished groundbreaking science by taking a novel approach to space exploration, achieving what has never been done before, and driving new technological advances that may benefit life on Earth and other stars and planets.

As a result, NASA intends to send humans to Mars by 2039, and Space X wants to go even faster, with plans to take people there by 2030. Mars' atmosphere is mainly carbon dioxide, the planet's surface is too frigid for human existence, and the planet's gravity is just 38% that of Earth.

Furthermore, the atmosphere on Mars is equivalent to about 1% of the atmosphere on Earth at sea level. This makes reaching the surface difficult. How will NASA arrive? How can we hope to survive in the face of such conditions?

When referring to human survival in space, it all comes down to Space Medicine. Space Medicine is critical to human space travel. It promotes survival, function, and performance in this difficult and potentially fatal environment.

Astromedicine: redefining Space Medicine

It is international, intercultural, and interdisciplinary, spanning exploration, science, technology, and medicine.

Space Medicine is also the most recent specialty recognized by the Royal College of Physicians and the General Medical Council in the United Kingdom. The curriculum explains the many types of spaceflights, environmental obstacles, associated medical and physiological impacts, and operational medical issues. It will detail the various tasks of the Space Medicine doctor, including surgical and anesthetic procedures, and will conclude with a picture of the future of Space Medicine in the UK.

Doctors in Space Medicine are in charge of space workers and spaceflight participants. 'Flight surgeons' and 'Space physicians,' with the assistance of other specialists, have created mitigation techniques to protect the safety, health, and performance of space passengers in an extreme and hazardous environment.

This comprises all aspects, from selection

to training and spaceflight, as well as post-flight rehabilitation and long-term health. The specialty's recent designation provides a road to training in this exciting field of medicine and is a critical enabler for the UK Government's commercial spaceflight aspirations.

We have seen that as space technologies and businesses have grown, Space Medicine has become increasingly neglected. To close the gap, specialists in Space Medicine, space technologies, and corporations interested in space exploration must collaborate.

More advancements in Space Medicine are required; flight surgery and research are insufficient to send astronauts into space for near future missions. This would jeopardize the health of the entire crew.

Why hasn't more research been done on the subject, one wonders? Why hasn't there been any progress in this area? In my view, the slow growth of Space Medicine can be summed up as a lack of communication and numerous agencies refusing to cooperate for political and economic reasons. All this eventually led to a chasm.

The bottom line is that we are not prepared for future missions. Thus, to prevent this, the global space effort will need to define a new model for what we currently call Space Medicine. And that model is called the ***Astromedicine Architecture.***

This book will attempt to address three overarching goals:
- Validate the need for change in this important area of research and development.
- Introduce the concept of ***Astromedicine*** as a model / architecture / framework for consideration to address the needed change.
- Clarify the model / architecture / framework in a manner that helps you, the reader, clearly understand how you can become part of solution and help humanity safely reach the heavens.

So, the goal of this book is to develop a new model of Space Medicine known as

V. Farsadaki

Astromedicine.

In terms of education, it would aid in the preparation of a new generation of workers capable of making Space a safe place for humans and a place not only to sustain humanity, but also to enable it to thrive.

The second goal of this book is to prove the economic reasoning to develop ***Astromedicine***, which will also sustain it, in the long run. It will investigate funding and investments for entrepreneurs with brilliant ideas for Space exploration and ***Astromedicine***.

I intend to fill the void that we are experiencing. One of our liabilities is the gap between technology and the missing medical elements. Our times aren't just about developing launch capabilities, but also propelling humans into space, and assuring their safety. In practice, we are not prepared for that. I propose a new model for the problems we confront, as well as potential answers.

So, what qualifies me to provide such

Astromedicine: redefining Space Medicine

ideas? Let me begin by explaining my educational and professional experience to assist you understand how I might provide potential answers.

I began my academic education with a bachelor's degree in biology from the University of Geneva. Following my BSc, I pursued my master's degree in Genetics at the same university. My desire to learn more did not end there, as a result, I moved to Italy to study Medicine at the University of Pavia. At the same time as Medical School I decided I needed to better understand the conditions of Space, so I simultaneously earned a master's degree in Astronomy and Astrophysics from the International University of Valencia. For most people, these degrees would suffice, but not for me; I enrolled in the MBA program at Quantic Business School of Technology shortly after obtaining my medical degree. Most recently, I had the privileged of graduating with distinction from the Thunderbird School of Management at

Arizona State University with an executive Master's in Global Management with a focus in Space Leadership. This was an excellent program under the leadership of Dr Greg Autry and Dr Zaheer Ali.

You're probably wondering what I was doing with all these degrees. Some have even told me that these degrees have nothing in common. Listening to these queries makes me giggle on the inside every time. I don't do anything without a plan or a goal in mind. There was always a plan. I've always wanted to be a **'Space Doctor.'** Not a flight surgeon, but a space doctor.

Aerospace physicians or space doctors provide for pilots, aircrews, and astronauts' health, safety, and well- being. Space Medicine, which has five recognized residency programs in the United States, is an appealing career option for physicians looking to conduct preventative or occupational Medicine with unique obstacles.

When preparing for a mission, flight

Astromedicine: redefining Space Medicine

surgeons are in charge of health care and medical training. They also handle any medical difficulties that may emerge prior to, during, or after a spaceflight. This is performed through a focused search for problems physical, emotional, environmental, and man made problems that jeopardize air and workplace safety.

To become a space doctor, I had to study a variety of fields, each with its own set of advantages. Every degree I've obtained to date is an important part of the larger process of becoming a space doctor. To learn about the conditions of the universe, I studied Astronomy and Astrophysics. We can grasp where we came from, where we're heading, and how Physics works in settings that we can't replicate on Earth. The Universe is our Astrophysics lab!

Cell and molecular biology research span all scientific disciplines in Space Biology, from understanding how single-celled organisms like protozoa, bacteria, and fungi respond to spaceflight conditions to how all of the cells in a complex tissue or organ work together to help an organism as a whole acclimate to such a foreign environment.

The primary purpose of NASA's Space Biology Cell and Molecular Biology study is to establish how the challenges of spaceflight affect living systems at the most fundamental cellular and molecular levels, employing cutting-edge cell and Molecular Biology tools and measurements. This includes recognizing and describing changes in gene and protein expression, DNA function and structure, cellular structure and morphology, and cell-to-cell communication.

Putting all of this together, I have a good understanding including Space conditions as well as human body from a clinical to a DNA level. After obtaining all my degrees, one can argue on whether I can be called "*Space Medicine expert*" in the classical sense. However, my extensive knowledge of understanding the science behind every space mission, how the human body reacts in space, and so on, have definitely led me to be a

Astromedicine: redefining Space Medicine

Space Medicine Expert. Furthermore, when you consider the business aspect (as I mentioned above, I have also completed an MBA), it pretty much qualifies me for the entire procedure of becoming a Space Doctor.

Having worked as a financial analyst for a trading firm as well as having a Trading Degree, I can put everything together to figure out how to make investments. We need to invest money to bring those fantastic concepts to life in order to advance the concept of *Astromedicine*, but more about that later.

To become a Space Doctor, one should have a lot of varied experience. As a result, not only did I complete a postdoctoral research project on radiation protection of astronauts, but I also briefly worked for NASA. Currently, I am the President and CEO of Space Exploration Strategies. I founded a company and scaled it to a workforce of twelve people. My company has three major areas of focus.

- The first area of focus is on the Business

Development of small entrepreneurs looking to enter the Space sector.
- The second is to write Space Medicine proposals or solicitations for companies seeking funding for their ideas or looking to fund others.
- Thirdly, my company offers Space Health advising to businesses looking to design and build something that is friendly to the physiology and psyche of a human.

For example, the human factor has yet to be addressed. The 'Human factor' in space environments focuses on five characteristics of human performance that are strongly affected by changed gravity and other aspects of long-duration space missions.

THE REVOLUTION

Every couple of times per year, the Aerospace Medical Association (AsMA) hosts a medical conference to share new discoveries or concepts, during which it is discussed anything mission-related to Space or Aviation research. As a result, I attempt to attend as many conferences as possible to learn about new inventions and discoveries.

As the idea for *Astromedicine* was being born within me and to better understand the people stating to work within "Space Medicine", I recall visiting an AsMA conference in 2022 (I wasn't presenting at that particular one), with the sole purpose of asking the as many attendees as possible, two important questions: "What is Space Medicine?" and "Do you believe we are ready for what's coming?"

I polled about twenty people and received about twenty-five different definitions in response to the first question. Nobody could agree on a single definition, which led me to the conclusion that nobody ever agreed on what Space Medicine really is. One highly ranked individual from the private sector seized me up and with a scolding tone replied: "If you don't know the answer to what Space Medicine is, young lady, you are in the wrong room" and then proceeded to turn their back to have a conversation with somebody more "age-appropriate".

After some thought, I concluded that there was a lot of hatred from people who were so heavily committed in something that is so thirsty to evolve. They understand exactly why we require Space Medicine. These same folks, however, are the ones who refuse to see what must be done and how things need to change in order to effectively make Space safe for humans. For the second question, however, everyone seemed to agree that we are not prepared for the future. Every single person I asked, replied "No". Needless to say, I left that conference convinced that I needed to start to change something in the field.

To summarize, the purpose of this book is to walk you through the problem, what we know thus far and my suggested solution. This consists of introducing the new discipline of

Astromedicine: redefining Space Medicine

"*Astromedicine*" and to summarize the road map of how we are going to make Space safe for humans.

Above all, we are unprepared for the future and what is to come in terms of upcoming missions. We are forcing astronauts to go on "suicide missions" due to the slow advancement of Space Medicine. The responsible authorities and governmental organizations involved in Space missions should become more aware of the issue. The responsible Space agencies should do extensive research into the difficulties and offered solutions from the various sectors that want Space missions to succeed. They should understand that historically, Space Medicine was an excellent answer, if not the sole solution. However, it is now time to repurpose the historical solution and combine it with advanced technology and education to create a proposed solution, **Astromedicine**.

My goal is to draw the attention of higher authorities to this issue that has gone underfunded for years. It is time to combine Space technology and Medicine to achieve the full success of our future Space missions.

People should be introduced to the notion of ***Astromedicine***, including what it is and how it is going to take over the Space health industry. We have the opportunity, in the next one to two generations, to be able to build a whole framework to ensure the safety of humanity in Space.

Astromedicine: redefining Space Medicine

Table Of Contents

Acknowledgments III
Introduction IV
    The Revolution XVI
Chapter 1 1
    History 1
        The Moon Landing 4
        The Space Shuttle 6
        What is Space Medicine? 8
        History and Evolution of Space Medicine 10
        Hubertus Strughold coins the term "Space Medicine" 11
        Interesting fact 11
        Technological Advances in Space 13
        Why is Space Medicine necessary? ... 18
        Flight Surgeons 25
Chapter 2 30
    Astromedicine 30
        What exactly is Astromedicine? 31

What Is the Difference Between Space Medicine and *Astromedicine*? 34
How come the sector still doesn't know how to address basic problems? 35
What do PhD researchers offer? 39
MDs and PhDs in the Workplace 41

Chapter 3 43
Democratization 43
What is the relationship between democratization of Space technology and accessibility? 46
Space Travel 48
Why should human health and survival in space be prioritized? 52
What is the definition of space? 54
Making Space safe 56
Ordinary People Doing Extraordinary Things 58
Benefits of Space Democratization ... 61

Chapter 4 63
Astromedicine's First Pillar: Education 63

Chapter 5 76

Astromedicine's Second Pillar: The Wheel 76
What is the relationship between these pillars? 80
Conclusion 86
Author Information 90

CHAPTER 1

HISTORY

Humans have always looked up into the night sky, gazed at the stars, and fantasized about Space, since the dawn of time. They not only fantasized about it, but also attempted to make it a reality. Humans' insatiable curiosity about life beyond Earth has recently been quite fascinating. Many 'firsts' in Space exploration have occurred in just the last few decades.

At the end of the twentieth century, powerful rockets were developed to overcome gravity and reach the solar system's end, paving the way for Space exploration to become a reality. On December 17th, 1903, Orville and Wilbur Wright created history by successfully flying the world's first airplane. They accomplished this astounding achievement near Kitty Hawk, North Carolina, after years of tireless experimentation and hard effort.

The Wright brothers' flight lasted only 12 seconds and covered only 120 feet. Despite its brief duration, it was a colossal feat that would eventually alter the path of human history. The brothers had spent years developing a powered flying vehicle, researching bird flight, and experimenting with different designs and engines. Finally, they built a biplane glider powered by a 12-horsepower engine. The glider had a control mechanism that allowed the pilot to modify the pitch and roll of the wings, giving them better flight control. The Wright brothers' glider was a substantial improvement over earlier designs, and it was this innovation that enabled them to fly.

On that momentous day, Orville Wright piloted the plane, while Wilbur operated the launching mechanism from the ground. Orville immediately gained altitude as the plane lifted off, leveling off around 10 feet above the earth. Before landing safely, he made a few changes to the wing pitch and roll, completing the first

Astromedicine: redefining Space Medicine

successful powered flight in history.

The significance of the Wright brothers' accomplishment cannot be emphasized enough. Their pioneering spirit paved the door for modern aviation, resulting in numerous technological improvements and altering the world as we know it. Air travel improved in efficiency, speed, and safety, transforming transportation and allowing people to travel farther places in less time.

Aviation is now an essential aspect of modern life, linking people and cultures all over the world. It all began over a century ago with the Wright brothers' historic flight in Kitty Hawk, North Carolina. Their legacy lives on, inspiring future generations to aim for the stars and push the limits of human achievement.

Later in the 1930s and early 1940s, Nazi Germany grasped the endless potential of long-distance rockets as weapons. Following that, 200-mile-range V-2 missiles that arced 60 miles high over the English Channel at more than 3,500 mph struck London during World War II. Following

World War II, the United States and the Soviet Union developed their missile systems.

On October 4, 1957, the Soviet Union launched Sputnik 1, the first artificial satellite into space. On April 12, 1961, Russian Lt. Yuri Gagarin became the first person to orbit the Earth in Vostok 1. Gagarin flew for 108 minutes and reached a height of 327 kilometers (about 202 miles).

Explorer 1, the first US satellite, reached orbit on January 31, 1958. Alan Shepard became the first American in space in 1961. Later, on February 20, 1962, John Glenn made history by becoming the first American to orbit the Earth.

The Moon Landing

On November 14, 1969, Apollo 12 launched its second moon landing. President John F. Kennedy set the national goal of "landing a man on the Moon and safely returning him to Earth within a decade" in 1961. On July 20, 1969,

Astromedicine: redefining Space Medicine

astronaut Neil Armstrong took "one huge leap for mankind" by landing on the moon. Six Apollo missions were launched to examine the Moon between 1969 and 1972.

During the 1960s, before astronauts ever landed on the Moon, unmanned spacecraft took pictures and conducted investigations. By the early 1970s, the Mariner spacecraft was orbiting Mars and mapping its surface, and orbiting communications and navigation satellites were widely used. By the end of the decade, the Voyager spacecraft had returned detailed images of Jupiter and Saturn, as well as their rings and moons.

The Apollo Soyuz Test Project, the first multinational crewed space mission (including Americans and Russians), and Skylab, America's first space station, were two of the most significant successes in human spaceflight during the 1970s.

When television programming was transmitted over satellites in the 1980s, people could use their home dish antennas to receive satellite transmissions. Satellites discovered an ozone hole over Antarctica, tracked down forest fires, and captured photographs of the 1986 Chernobyl nuclear power plant disaster. The discovery of new stars by satellites gave us a new viewpoint on the galactic center.

The Space Shuttle

Following the launch of the Space Shuttle Columbia in April 1981, the reusable shuttle was used for the majority of commercial and government Space missions. Up to January 28, 1986, when the Space Shuttle Challenger exploded barely 73 seconds after takeoff, 24 successful shuttle launches met a variety of scientific and military requirements. Christa McAuliffe, a New Hampshire teacher who would have been the first civilian in space, was among the seven crew members killed in the tragedy.

The Space Shuttle was the first reusable

vehicle, carrying passengers into orbit, launching, recovering, and repairing satellites, conducting groundbreaking research, and assisting in the construction of the International Space Station.

The Columbia disaster was the second shuttle disaster. On February 1, 2003, the shuttle disintegrated upon re-entering the Earth's atmosphere, killing all seven crew members. The catastrophe occurred over Texas just minutes before it was planned to land at the Kennedy Space Center. An investigation determined that the accident was caused by a piece of foam insulation that became separated from the shuttle's propellant tank and injured the left wing's edge. It was the second time a shuttle had been lost in 113 shuttle flights. Following each disaster, space shuttle flight operations were halted for more than two years.

Discovery, the first of the three operational space shuttles, performed its final flight on March 9, 2011, with Endeavour following suit on June 1. The landing of Atlantis on July 21, 2011, signaled the conclusion of the 30-year space shuttle program and the final shuttle mission.

Space systems are most important in homeland defense, weather monitoring, communication, navigation, imaging, and remote sensing for chemicals, fires, and other disasters. Space Medicine, which should have been an intrinsic component of the process, has been neglected for as long as Space technology has grown, and all Space missions have been done for Space exploration.

What is Space Medicine?

Since the Wright brothers, Medicine had to adapt to those that flew pilots and eventually cabin crew too. However, soon humans started flying higher and higher until they surpassed the Karman line. Who had to take care of those patients? The same doctors that took care of pilots. The environment within the Earth's

Astromedicine: redefining Space Medicine

atmosphere and magnetosphere versus outside of those, are very different. Furthermore, the periods of time that pilots and astronauts were flying were also quite different from each other, with the first group to be above the Earth's surface for hours versus the second group to eventually be away from the ground's protection for months on end. Starting to see the problem?

Flight Surgery, a branch of Medicine, was officially founded in the 1950s to aid with human Space travel. It began by discussing how microgravity had an immediate impact on human physiology. As flight times increased, so did our awareness of the long-term effects of microgravity, radiation, and solitude, among others.

Space Medicine, like terrestrial Medicine, evolves progressively as spaceflight technology pushes the limits of our knowledge and bodies. The quick pace of Space activities at the dawn of the Space age left little time for the development of a medical foundation for the human Space program. Life support, safety, and health concepts came first, and they were primarily based on accepted Aviation Medicine.

Because the Space race was primarily motivated by political considerations, it was only natural for astronauts to be military test pilots, at the time. Scientists and medical professionals were eventually recruited on board to conduct research and contribute their expertise in the medical area. Story Musgrave was the first to be selected, and he flew six Space Shuttle missions between 1983 and 1996.

History and Evolution of Space Medicine

Space Medicine had to advance quickly in order to support the presence of people in Space an environment for which they were not designed. Evidence based Medicine, research learning, and clinical experience were the driving elements behind the never-ending effort to preserve astronauts' lives and understand how the Space environment impacts human physiology.

Hubertus Strughold coins the term "Space Medicine"

Hubertus Strughold coined the term "Space Medicine" in 1948. Strughold was a former Nazi physiologist and physician who came to the United States as part of Operation Paperclip after WWII, along with Werner von Braun and Arthur Rudolph. Strughold was named the first professor of Space Medicine at the School of Aviation Medicine. Later, he was named director of the Department of Space Medicine at Brooks Air Force Base in Texas, which is now known as the US Air Force School of Aerospace Medicine.

Interesting fact

Strughold, who was instrumental in

developing the pressure suit as well, used by the first US astronauts, lost most of his awards after being convicted of war crimes at the Dachau concentration camp.

Motion sickness was the first ailment to be treated, but as missions became longer, additional health concerns arose. Astronauts had to be alert and responsive in addition to being in good health. Dextroamphetamine sulfate, a powerful stimulant known as a "go-pill," was taken by astronaut Gordon Cooper on the final Mercury Project flight, the Mercury-Atlas 9.

Cooper was the first to use the pill in flight, despite the fact that it was available in both the spacesuit and the survival pack. Unfortunately, when the International Space Station (ISS) replaced the Gemini, Apollo, Skylab, and Space Shuttle missions as the primary means of long- duration Space travel, widespread illnesses followed. The medical package for each program was tailored to the crew size, anticipated activities, and mission duration.

When he wasn't in Space, Musgrave used his medical knowledge to help design medical equipment for the Space Shuttle program. All manned missions' medical packages have flown as clear examples of how Space Medicine has evolved from a few drugs to sophisticated diagnostic equipment and instruments to handle the majority of ailments that can occur while in orbit.

Technological Advances in Space

Emerging technologies such as 5G, advanced satellite systems, 3D printing, big data, and quantum computing are being used by the space industry to improve and scale activities in space. Weather forecasting, remote sensing, GPS navigation, satellite television, and long-distance communication all rely on Space infrastructure. Smart propulsion, Space robotics, and Space traffic management are among the new Space

Tech concepts gaining traction in the Space industry.

Startups develop technologies that facilitate travel, operations, and communications between Earth and Space, while also encouraging private investment in the industry.

Small satellites have become more common in recent years, making them the main Space Tech trend in 2023. Smaller spacecraft can be designed more economically, and advances in industrial technology enable mass production of these satellites. Startups develop small satellites to enable space firms to perform activities that large satellites would ordinarily find challenging. Small satellites are also excellent for private wireless communication networks, scientific observation, data collection, and GPS earth monitoring.

To improve goods and services, new technologies are being integrated into space manufacturing. With the development of powerful robots, 3D printing, and light-based production, the space sector is progressing. Large space structures, reusable launch vehicles, space shuttles, and satellite sensors are now a reality thanks to advancements in manufacturing techniques. Long-term space exploration and missions rely on automation, thus companies provide solutions tailored to the industry.

Space communications, at their most basic, rely on a transmitter and receiver. A transmitter encodes electromagnetic waves with a message. These waves then flow through space as they approach the receiver. Recent advancements in space communication, on the other hand, go beyond simple transmitters and receivers to provide advanced space communication via large-area antennae, base stations, and LEO satellites.

The vast majority of LEO spacecraft constructed by mankind are now space junk. This includes abandoned satellites, rocket engines, and, most crucially, minute particles of debris from collisions and explosions. Because of all of this garbage, the future of space exploration and travel is jeopardized. Startups are developing practical solutions for debris retrieval and space traffic control to address this issue.

Satellite constellations rely heavily on their in- space propulsion technology. Given the costs and environmental implications of space missions, businesses are looking for solutions to ensure the sustainability of these journeys.

As a result, multinational startups and scale-ups are investigating many solutions, ranging from electric, green, and water based propulsion to iodine-based propulsion systems, to enable the next generation of clean rockets in space.

Space movement and activity control is a growing trend in space technology. Some examples of space activities include tourism, commercial space missions, satellite maintenance, food production, garbage removal, and space station modifications. Such advancements give the scientific community more leeway by

Astromedicine: redefining Space Medicine

allowing them to study how living things behave in space.

Space exploration provides essential answers to fundamental questions concerning the nature of the universe and the evolution of our solar system. By tackling the issues involved with space exploration, humans discover opportunities to further mining, material science, and life science research. Furthermore, space missions push the boundaries of science and technology while inspiring the next generation of students, educators, and researchers around the world.

When mining celestial bodies, science fiction (Sci-Fi) gives way to reality. Asteroid mining by private persons and businesses employing advancements in space cameras and satellites allows for the precise location of asteroids. If found, these heavenly planets can be used to extract minerals such as platinum, gold, iron, or even water. There is an obvious economic motivation for space mining, and analysts say it has the potential to become a

Multibillion dollar industry.

Why is Space Medicine necessary?

The advancement of cutting-edge technologies has enabled previously unreachable ambitions to be realized and has broadened the scope of what the human species is capable of. The hazards that astronauts may face are severe, and they could have a significant impact on their bodily and mental well-being. Specific symptoms of emotional dysregulation, cognitive dysfunction, disruption of sleep wake rhythms, visual phenomena, major body weight changes, and morphological brain abnormalities are among the most frequently reported events during Space missions.

Astronauts' sleep patterns are considerably altered during spaceflight, according to Dr. Andrew Liu, an MIT research expert.

"There will not necessarily be the same

Astromedicine: redefining Space Medicine

24-hour clock you receive here on Earth after you fly up into space," Liu explains. As a result, in its absence, you must either (make an artificial sleep schedule) or discover a way to monitor and ensure that folks get enough sleep. Chronobiology is thus a branch of Biology that studies timing mechanisms, such as periodic (cycle) occurrences in living organisms, such as their adaptation to solar and lunar rhythms.

Because of their orbit around the planet, astronauts aboard the International Space Station (ISS) see 16 sunrises and 16 sunsets during each 24-hour day. The artificial light in the space station considerably leads to astronaut sleep cycle disruption and a major shift in dark- light cycles. According to Liu, the majority of the artificial lighting on the station is made up of a certain wavelength of blue light.

The astronauts were unlucky because 'when you are exposed to blue light late at night, your body starts thinking, 'Well, you want to stay up later,' so it attempts to keep you awake, so you do not receive the release of melatonin that helps you sleep.

Individuals who regularly experience sleep loss have a lower executive function or inhibited mental operations, such as concentration and decision-making, according to Liu. People who routinely sleep less than they

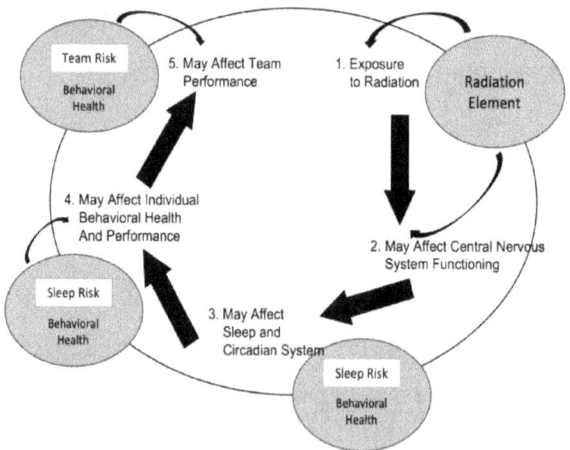

need tend to perform worse on mental tests, even if they believe they have adjusted to the lack of sleep after a while.

Note: This is an adaptation of Landon et al.'s 2016. Example of possible Behavioral Health risks arising from Space Environmental

Astromedicine: redefining Space Medicine

Conditions.

A disturbed sleep cycle may be damaging to astronauts' tasks, which might range from conducting experiments to mending satellites, in addition to potentially hurting their mental health. Valentin Lebedev, for example, had an inconsistent sleep cycle while suffering from depression (Potter, 2008).

The diagram above shows how environmental dangers can lead to mental health difficulties, which can then lead to problems with team dynamics.

A sleep-deprived astronaut in command of a spacecraft may jeopardize their crewmates. An astronaut with a poor sleep cycle may also endanger the other members of the team.

According to Dr. Moore, radiation exposure is most certainly an aspect of space travel that can have a harmful impact on an astronaut's mental health. The majority of the harmful radiation emitted by our Sun, our galaxy and the Universe is shielded by the Earth's atmosphere and magnetosphere, keeping as such humans safe. Astronauts, on the other hand, are more likely to be exposed to all ionizing radiation types while in Space, which can cause radiation sickness, nervous system damage, increase the risk of acquiring cancer and even create mutations at a genetic level, dangerous to pass on to their children.

According to Moore, astronauts "seeing 'flashes' while going to sleep" are most likely the result of particles "tearing through" their eyes and brains (Moore, 2018). According to Liu, one of the least understood risks associated with longer missions is increased radiation exposure: "Radiation is the one wild card factor. You could get a fair sense of how people would feel about being on that mission and limiting their exposure to [blue] light [for sleep] if there was no radiation harm."

Because journeys to Mars and other planets would last much longer than current space missions, the effects of radiation on astronauts must be better understood.

Mission Type	A	B	C	D
Duration (up to)	6 Months	12 Months	12 Months	12-36 Months
Distance from Earth	Low Earth Orbit	Low Earth Orbit	Deep space Exploration	Deep space Exploration
Crew Size	6	6	4	4-6
Vehicle Size	Large	Large	Medium/ Small	Medium/ Small
Communications delay (one-way)	0.5-3 seconds	0.5-3 seconds	8-10 Minutes	10-20 Minutes

Note: This is an adaptation of Barret et al.'s 2015. Mission profiles for astronaut job analysis.

Astronaut John Grunsfield's work on the Hubble Space Telescope Project is just one example of the many difficult tasks assigned to astronauts (NASA, 2009).

Crew members must work hard on activities such as fixing the ISS, while they are constantly monitored by experts on Earth. Living in a small area and being away from their family for extended periods of time is a difficult job. In the 1980s, Chinese payload specialist Taylor Wang vowed not to return to Earth after his experiment went wrong due to the strain and constant pressure.

"We had one payload specialist who became obsessed with the hatch. 'You mean all I have to do is turn that handle, the hatch opens, and all the air goes out?' It was scary...so we began to lock the hatch," astronaut Harry Hartsfield remarked of a crew mate (Morris, 2017).

The Soviet Soyuz 21 mission was forced to terminate in 1976 after the crew complained of an unpleasant odor; it is possible that the entire crew experienced the same delusion. It is interesting to note that the source of the odor was never identified, and a recent NASA assessment suggested that the crew may have imagined the odor (NASA, 2016b).

There were no recorded mental health issues until the 1990s, and many of the psychological problems encountered by the crews of the Mir and Salyut missions may have been caused by "earthbound concerns," such as having crew members with divergent personality types in a small environment (Inglis-Arkell, 2012).

According to Beven, some NASA astronauts on the Mir Space Station experienced sadness and loneliness (Inglis-Arkell, 2012). For example, the Soyuz 21 trip to the Salyut 5 station, which was canceled after the crew complained of a foul smell, was said to have experienced high tension between the cosmonauts (NASA, 2016b).

Flight Surgeons

Given the enormous investment in training and mission preparation, flight crew personnel must be kept as healthy as possible. Returning to Earth from the mission is as important and dangerous as going to space. To that end, NASA assigns a unique physician called a "Flight Surgeon" after a crew is assigned to a mission. Flight surgeons handle any medical difficulties that arise before, during, or after spaceflight, as well as the health care and medical training astronauts get in preparation for their mission.

Flight surgeons attend weekly confidential medical briefings with the astronauts while operating the console at NASA Mission Control Center (MCC) under the call sign "Surgeon" during a mission. Flight surgeons do not relay conference specifics to Mission Control, but they do notify the flight director if they discover anything that could jeopardize the mission.

While the definition of Flight Surgeons mentioned above is quite formal, there are two assumed groups. Flight Surgeons (aka specialized MDs) are in direct contact with the astronauts, dealing with their labs, assisting and preparing them for the mission, and ensuring they can travel in space. These flight surgeons are convinced they are the only ones capable of

Astromedicine: redefining Space Medicine

practicing Space Medicine.

On the other hand, there is the group of PhDs/ researchers in academia and industry that claim to be working on the true essence of Space Medicine, as they are the ones answering basic questions like, "What happens to liver cells if exposed to X amount of ionizing radiation?" or "What happens to metabolic pathway Z after Y amount of time in microgravity?" These questions are critical for us before embarking on a mission. We were thirsty for this kind of evolution, and we got it after 60-70 years of Space exploration.

Furthermore, the Engineers who have been designing and building the vehicles for astronauts and their missions, as well as their flight suits were also convinced that their aspect of Space Medicine is closest to the "real definition". Interestingly enough, they did not consider the consequences that tiny cubicles could have on the astronauts' mental and physical health. Ideally, Aerospace Engineers and Flight

Surgeons who claim to study Space Medicine should collaborate and analyze every aspect of the technology sent into Space and its effects on human health.

Another group of people that has been playing a key role in Space Medicine and its evolution is that of those who "put the money into the pot". These have traditionally mainly been politicians who approved budgets and bills, however there were always the private investors too. Nowadays, investors might want to invest in Space health but not know where to start or how to go about it. Which ideas are worth investing?

The biggest issue in all this? There is a lack of communication between these four groups. We have mentioned only four here, but there is an extensive list of key players in what has been called up until today "Space Medicine".

The evolution of Space Medicine has been exceedingly poor, with most mission conductors ignoring even minor countermeasures. Now, if we talk about NASA's planned 2039

Astromedicine: redefining Space Medicine

Mars Mission, it can only be successful if we take countermeasures and do not send astronauts on a death mission.

Deep space travel is currently consisted of "suicide missions" not due to the doubt of whether the crew will make it back to Earth – we are quite ready for this, as Space technology can ensure their safe return, but rather on what kind of harmful traits the trip will imprint on their DNA.

We can't make any predictions because we haven't done enough research on the subject. It is easy for anyone to see that the successful and fast evolution of the Space race from the '60s was driven by political and economic reasons. While nowadays, these reasons might still drive technological advancements, it is undeniable that the private sector has revolutionized Space technology. So, I ask you: Who will revolutionize Space Medicine?

CHAPTER 2

ASTROMEDICINE

For many decades, people around the world have been captivated by the prospect of humans venturing into the unknown reaches of Space, from the historic moon landing to the more recent Space X missions. However, while the focus has primarily been on Space Technology and Engineering, the importance of Space Medicine has often been overlooked.

The reality is that Space exploration involves two equally important aspects: Space Technology and Space Medicine. While Space Technology is critical in enabling human Space travel, the sustainability of human life in Space ultimately depends on advancements in Space Medicine. Human spaceflight would not be possible without proper medical support and care. Consider the following scenario: we successfully

Astromedicine: redefining Space Medicine

launch a spacecraft into deep space, but the crew becomes ill due to the harsh Space environment or suffers an injury, and we are unable to safely return them to Earth; this would be a devastating and costly failure. It is therefore critical to ensure the safety and well being of astronauts in Space.

The interpretation of Space Medicine can vary among experts in the field, ranging from Flight Surgeons, to Researchers, and from Engineers to Business Investors. Even though most people of the field can agree that Space Medicine is about humans in Space, the details of what the discipline really comprises tends to be disputed.

What exactly is Astromedicine?

Astromedicine is a multidisciplinary field of study that aims to ensure the safety and well-being of humans in space. It includes aspects of research, technological engineering, medical practices, and their respective sub specialties, as well as business and investment aspects to facilitate the necessary funding and resources for this revolutionary field of study.

The significance of *Astromedicine* cannot be overstated, as it has become increasingly clear that Space is the future of human exploration and colonization.

However, Space is an inhospitable environment that poses numerous human health and safety challenges; human Space exploration would be impossible without adequate measures to mitigate these risks.

Astromedicine aims to address these challenges by researching the effects of microgravity, radiation exposure, and other space-related factors on human health. This research is used to develop and implement advanced technologies and medical practices that can protect human health in Space.

Astromedicine's business and financial parts are equally critical, as they provide the required financing and resources to expand this

Astromedicine: redefining Space Medicine

field of study, such as securing funding for research and development and investing in companies specializing in Space medical technology and procedures.

Due to the need for a more comprehensive and unified approach to Space Medicine, I coined the term "*Astromedicine*", a new form of Space Medicine Space Medicine 2.0. The lack of consistency in defining Space Medicine by different teams involved in Space exploration created gaps in our understanding of the field. Additionally, while the term "Space Medicine" was well-known, it had become limited in scope and failed to convey the full complexity of medical needs in space.

As a result, I believe that a new term was required to capture the breadth and depth of medical considerations in space. "*Astromedicine*" represents a new approach that encompasses all aspects of medical care in Space and allows us to break free from the constraints of the previous definition; it was a necessary step toward creating a road map for the future of Space exploration and Medicine.

What Is the Difference Between Space Medicine and Astromedicine?

While it is not entirely true that every team in Space Medicine works independently without collaboration, it is true that there is a significant lack of collaboration and communication in the field.

Space Medicine may in theory involve a collaborative effort between various teams of medical professionals, engineers, and scientists who work together to address the unique health challenges that arise in Space, such as Flight Surgeons, Space physiologists, Space psychologists, and other experts who collaborate to develop and implement medical protocols, technology, and strategies to support astronaut health and safety, however the hard truth is that these groups of people focus on their field and

Astromedicine: redefining Space Medicine

rarely ask for input or even discuss issues they might be facing with others.

Astromedicine, on the other hand, can be a more integrated approach to Space exploration, in which medical professionals, engineers, scientists, and other experts collaborate to develop solutions, such as designing spacecraft and habitats, developing medical protocols and equipment, and developing strategies for managing emergencies and maintaining crew members' health.

In summary, while Space Medicine has ended up dividing its members, **Astromedicine** will take a more comprehensive and integrated approach to making Space safe for humans, incorporating all relevant fields in partnership.

How come the sector still doesn't know how to address basic problems?

There are distinct groups responsible for ensuring the health and safety of astronauts, each with their own approach to Space Medicine, as discussed earlier. However, Flight Surgeons, who are the classical experts in Space Medicine, could have had the potential to perform the duties of all these groups with the correctly adjusted training and resources. Becoming a Flight Surgeon is no easy feat, as it requires multiple residencies after medical school and extensive training alongside astronauts.

    They generally train in Space-related emergency care and internal medicine, but their curriculum doesn't even mention the physics of being Space. One would think that comprehending ionizing radiation for example or even the way that a simple proton can break a double strand of DNA, should be essential. However, these aspects are not covered neither in the training nor the education aspect of Flight Surgery residency.

    Despite being regarded as experts in Space Medicine, Flight Surgeons may be unprepared to handle certain conditions in Space

Astromedicine: redefining Space Medicine

due to a lack of knowledge and training, which is a significant issue that limits their ability to perform their duties effectively and fully cater to the needs of the astronauts.

One of the most serious problems astronauts encounter is radiation. The vastness of space has enticed many daring explorers, but with the exception of the Apollo astronauts, all others have remained confined to Low Earth Orbit, which is shielded by the Earth's magnetic field and atmosphere, and venturing beyond this orbit would mean exposing oneself to solar radiation and hazardous cosmic rays.

Beyond the protective shield of the Earth's magnetic field lies a universe teeming with radiation unlike anything we've ever seen before, including high speed solar particles from the Sun and galactic cosmic rays from beyond our solar system.

The Earth's atmosphere and magnetosphere provide a protective shield from harmful radiation. The Earth's atmosphere consists of four primary layers: the troposphere, stratosphere, mesosphere, and thermosphere, which absorb harmful radiation to keep us safe on the surface. While our genetic evolution did not provide us with inherent radiation protection, we are shielded as long as we remain on the Earth's surface.

However, when astronauts travel into space, they are exposed to the sun's harmful rays, which can cause significant harm. To protect against radiation, spacesuits include protective layers that reflect and absorb incoming rays, as well as a gold-plated visor section to shield the astronaut's eyes.

Furthermore, modern spacecrafts have multiple bumper shields made of thin aluminum sheets, a net made of Kevlar and epoxy (high in hydrogen and used in military and firefighting gear), and air gaps between them to slow down radiation particles.

When an incoming solar flare wave of protons is detected on the International Space

Astromedicine: redefining Space Medicine

Station (ISS), a system alerts the crew with an alarm, and the crew has only a few minutes to retreat to their designated safe quarters, aka bedroom capsules. The purpose of taking refuge in these quarters is to reduce the potential damage from the incoming wave of protons.

The quantity of radiation remains constant, so there is no prevention of radiation exposure. The described process is a prime example of the complexity of the radiation environment in space and the challenges of protecting human space travelers from its harmful effects. It also serves as a reminder that even experts in the field may not fully understand all aspects of radiation exposure and its effects on the human body.

What do PhD researchers offer?

I worked in research laboratories focused on genetic radiation protection, where we did the groundwork, so that one day we can design medications that over express genes coding for radio-susceptible proteins or similarly under express genes coding for radio exposing proteins from an astronaut's DNA to protect them during the space flight. This was an fascinating experiment that I wanted to look into, since I can remember my basic understanding of genetics.

Blood samples are taken before, during, and after the mission to measure white blood cell numbers and gene expression levels, so as to identify the release of toxic elements caused by radiation exposure. Researchers are constantly striving to find solutions to protect astronauts from radioactive waves in space. Currently, no preventive measures are available, and astronauts can only take cover in their chambers when alerted of incoming solar flares.

This is just one of the long list of fields that Space- related bench work research is examining. To make Space safe for humans, a combination of expertise from researchers and medical professionals is required.

Astromedicine can significantly

contribute to this goal, not just as a term, but as a revolutionary approach to the field.

MDs and PhDs in the Workplace

It is abundantly evident that a major issue must be addressed, a communication gap between every organization and field trying to create a sustainable and efficient work environment for humanity. This is where **Astromedicine** bridges the gap and protects the safety of human existence in space.

The importance of this concept extends beyond flight surgeons and researchers, as engineers who build spaceships and spacesuits play an important role in making space safe for humans. Each individual, regardless of their field of work, has a unique contribution to make, and it is only by working together that we will be able to find solutions to the problems faced by crew members in space.

We must recognize the diversity of human

anatomy and needs, which is where personalization within the spacecraft comes in. Engineers must create stimulating environments to combat the boredom and other mental health issues that crew members are bound to experience during long space journeys. We must also explore business and investment concepts that can help monetize and advance the field of ***Astromedicine***.

We can push the boundaries of Space Medicine and make space travel a reality for many by developing new and innovative ideas. My goal is to unify different fields and establish them as one concept - ***Astromedicine***. Let us restore the potential of Space Medicine and pave the way for a safer and more exciting future in space.

CHAPTER 3

DEMOCRATIZATION

The term *"democratization"* refers to making anything more generally available to the general populace, typically implies power or access, and is derived from the word "democracy," which refers to a type of governance in which the people or their chosen representatives have power.

The term 'democracy' derives from the Greek language, where it combines two shorter words: 'demos,' which means whole citizens living inside a specific city-state, and 'Kratos,' which means power or rule.

Democratization can be applied to numerous elements of society, such as politics, education, and technology, with the goal of increasing equality and participation among persons or groups who were previously marginalized or excluded from these areas.

Similarly, space democratization refers to making space exploration and utilization more accessible to everyone, rather than just government agencies or wealthy private companies, and it entails creating opportunities for more people to participate in space- related activities such as scientific research, space tourism, and even living and working in space.

This can be accomplished through initiatives such as public-private partnerships, crowdfunding, and the development of reusable rockets and spacecrafts that reduce the cost of access to space. By democratizing Space, we can make Space exploration and utilization more inclusive and diverse, thereby accelerating progress toward a better understanding of our universe and our place in it.

Furthermore, we should emphasize the importance of equality and equity in Space, which means providing equal but appropriate opportunities for all. Although increasing public access to Space and Space-related technologies appears complex, it has gained popularity in recent years due to a number of factors.

One reason for this is the emergence of private Space companies such as Space X, Blue Origin, and Virgin Galactic, which have made significant advances in developing reusable rockets and spacecraft, disrupting the traditional government-dominated Space industry and lowering the cost of access to Space.

Another reason is the general public's growing awareness and interest in Space exploration and research, which has been fueled by technological advancements, the popularization of Space-related media, and the growing role of Space in addressing global challenges such as climate change and resource depletion.

Furthermore, Space democratization is seen as a way to promote innovation, entrepreneurship, and diversity in the Space industry, as well as an inspiration for the next generation of Space scientists and engineers.

Overall, the concept of Space democratization is a fascinating one because it signifies a huge transformation in how we view and use Space, from a realm controlled for governments and elite organizations to one open to all.

What is the relationship between democratization of Space technology and accessibility?

It refers to increasing the accessibility of Space technology and resources to a broader variety of people and organizations.

Space exploration has traditionally been limited to a few nations and organizations with significant financial and technical resources; however, with the increasing availability of low-cost technology and the rise of private Space companies, the barriers to entry in the Space industry are gradually lowering.

As the cost of Space technology and resources falls, more people and organizations

Astromedicine: redefining Space Medicine

will have access to Space exploration and usage, opening up new avenues for innovation and growth in scientific, technology, and engineering sectors.

Furthermore, by providing new tools for monitoring and understanding our planet, the democratization of space technology and resources can help to address some of the world's most pressing challenges, such as climate change. As an example:

Consider a crew of astronauts on a mission to Mars; their spacecraft has advanced propulsion engines, navigation tools, and life support systems that allow them to travel to Mars safely and effectively.

They employ powerful sensors and imaging technologies as they approach Mars to scan the surface and locate suitable landing spots, as well as real-time weather data and satellite imagery to assist them plan their descent and landing.

Once on Mars, the astronauts use

advanced suits and equipment to protect them from the harsh Martian environment, allowing them to safely explore the planet's surface and conduct scientific research, as well as advanced communication systems to stay in touch with mission control back on Earth.

All of these advancements in Space technology allow astronauts to travel to and explore Mars with relative ease and safety, despite Space travel's many challenges and dangers. As a result, we need to develop advanced Space technology so that traveling into Space is easier and safer rather than a complicated process.

Space Travel

Space tourism is a type of commercial space travel in which people pay to travel into Space for recreational purposes. While this business is still in its early phases, several companies are creating space tourism programs

Astromedicine: redefining Space Medicine

that will allow ordinary citizens to experience space travel.

Now, let's take a moment as a side note to mention that as people boarding a aircraft are called "Passengers" and not "Pilots", as such people boarding a spacecraft should be called ***"Spacengers"*** and not ***"Astronauts"***. Despite whether that might frustrate you, we have to agree on some terms sooner than later and it is widely known that the 4 minute requirements are not comparable to the Apollo 13 mission's requirements.

Space tourism aims to democratize access to space by making Space more accessible and affordable to the general public. By allowing ordinary citizens to experience Space travel, Space tourism can help create a greater interest in Space exploration and encourage more people to pursue careers in Space-related fields.

Furthermore, the revenue generated by Space tourism can be used to fund additional Space technology research and development, which can help advance our understanding of the Universe and potentially lead to new discoveries and innovations on Earth, too.

Space tourism is one method that Space democratization is achieved by making Space more accessible and inexpensive to a larger range of individuals, hence increasing public interest in Space and improving Space exploration and technology. A few examples:

Blue Origin's New Shepard spacecraft: In 2015, Blue Origin announced the development of its New Shepard spacecraft, which was named after astronaut Alan Shepard, the first American to go to space. In 2018, the New Shepard made its first successful crewed flight, carrying Blue Origin employees into space.

Virgin Galactic's Space Ship Two: In 2010, Virgin Galactic unveiled its spacecraft, which was designed to be launched from a carrier aircraft at a high altitude, then ignite its rocket engines to climb to the edge of space. In 2019, Virgin Galactic completed its first crewed spaceflight, with its founder Richard Branson among the onboard passengers.

Blue Origin's lunar lander: In 2019, Blue Origin revealed plans to build the Blue Moon lunar lander, which will bring cargo and, eventually, astronauts to the moon, with the goal of establishing a permanent human presence on the moon.

Virgin Galactic's Mach 3 aircraft: In 2020, Virgin Galactic announced plans to develop a Mach 3 aircraft capable of flying at three times the speed of sound, which would be used to transport passengers around the world in a matter of hours and could potentially launch small satellites into space.

These advancements were originally considered science fiction concepts, but they are now a reality, demonstrating the remarkable progress that has been made in the field of space tourism in recent years, with only a substantial sum required to purchase a ticket and fly to space.

V. Farsadaki

"The democratization of Space demands the democratization of Space health."

Why should human health and survival in space be prioritized?

When we talk about the democratization of Space, we mean making it accessible to all humans, which means having access to resources for energy, technology, asteroid mining, and other numerous benefits. However, the most important section has been neglected for years now - the health and survival of humans in Space.

Prioritizing human health and survival in Space are critical for several reasons. For starters, Space is an extremely hostile environment in which humans face a variety of physical and psychological challenges, such as radiation exposure, microgravity, and isolation. Maintaining good health and well-being is therefore critical for ensuring that astronauts can

Astromedicine: redefining Space Medicine

carry out their missions safely and effectively.

Assume you are an astronaut on a long-duration mission to Mars. During the journey, you are exposed to high levels of radiation, which may increase your risk of cancer and other health problems. Additionally, the microgravity environment can cause bone and muscle loss, limiting your ability to perform essential tasks upon arrival at Mars.

Prioritizing your health and mitigating these risks, such as exercising regularly and using radiation shielding, will better equip you to carry out your mission successfully and return to Earth in good health. Finally, prioritizing human health and survival in Space is critical for enabling us to explore and learn about the universe while ensuring the safety and well-being of the astronauts who venture into the final frontier.

Before we go into the overall concept of Space democratization, let us first explore the roots: What does "making space safe for humans" really mean?

V. Farsadaki

What is the definition of space?

We frequently use the single term "Space" to describe our expanding universe, but where does it begin and what is it?

Space is almost completely devoid of matter, has minimal pressure, and is almost a perfect vacuum; sound cannot travel through Space due to a lack of molecules positioned close enough to one another. Bits of gas, dust, and other matter float in the "emptier" (but not quite empty) parts of the cosmos, while planets, stars, and galaxies can be found in the more crowded parts.

The Kármán line, about 62 miles (100 kilometers) above sea level, is where we on Earth typically think deep Space begins. This is an arbitrary line at an altitude where there isn't much air to breathe or light to disperse, and beyond this height, blue begins to give way to black because there aren't enough oxygen molecules to keep the sky blue.

Astromedicine: redefining Space Medicine

Because of what we can see in our detectors, the precise scale of space is unclear; we use "Light-years" to quantify huge spans of space; one "Light-year" is equal to approximately 5.8 trillion miles (9.3 trillion kilometers). We have mapped galaxies almost as far back as the Big Bang, which is thought to have started our Universe about 13.8 billion years ago, using visible light in our telescopes, implying that we can "see" into Space at a distance of almost 13.8 billion light-years. However, as the cosmos expands, it becomes more difficult to **"Measure Space."**

Astronomers also dispute on whether our globe is the only one, implying that Space may be far larger than we currently believe.

However, the democratization of Space is dependent on how we define space: is it reaching the International Space Station (ISS), the Moon, Cislunar, Mars, or beyond? Or is it reaching a free-floating space settlement?

V. Farsadaki

Making Space safe

The definition of safety in Space can range from simply surviving and returning safely to Earth to thriving in Space with advanced technology and life support. The limits of democratization in Space will need to be set or at least defined, taking into account the physical and mental conditions of all humans who may want to go to space.

The concept of humans in Space is diverse and cannot fit into a box, including those with disabilities, diabetes, anxiety, depression, and varying body shapes and sizes. As a result, any discussion about democratization in Space must consider every aspect, including the destination and level of safety.

When discussing safety, one end of the spectrum is survival, which is relatively straightforward. However, five key pillars at NASA are currently receiving funding and attention, particularly life support and health. One of these pillars is radiation protection, which is critical for the 650 astronauts who have flown into space thus far.

On the other end of the range is thriving, which we have already discussed, such as the prospect of conception in Space, which raises many issues.

For example, if we travel to Mars for six and a half months, boredom will most likely be a major issue; during this time, people may forget to take their medication and require stimulation to combat boredom. Just as an example, there is also the possibility of pregnancy; who would be responsible for deciding whether to keep the baby? The mother's country? The father's State of origin? Or should it rather be an international affair?

If we decide to keep the baby, do we have the necessary medical equipment to support pregnancy and childbirth in Space? It took humanity a long time to eliminate death during labor on Earth. Imagine how much more complicated it will be in microgravity. If we decide not to keep the baby, do we have the technology and medical equipment to safely terminate the pregnancy?

This shift would be part of the democratization of Space since we would have developed a person capable of surviving and thriving in space. All these considerations are classified as safety concerns. This situation has Space Bioethics written all over it, as the first breath taken by a baby born in Space might expand their lungs in such a way that coming back to Earth, they might collapse, creating thus, a human who can't live on Earth: the first homo spaciens.

Ordinary People Doing Extraordinary Things

Most individuals are not trained like astronauts since it is more than just meeting physical and mental health criteria. Astronauts go through intensive training that can span anywhere from 2 to 10 years, with an average of 5-10 years, and includes significant physical and psychological pressures.

People who want to travel to Space, on the other hand, cannot (and should not have to) access this training, raising the question of what technology and support are required to make Space accessible to everyone.

Making Space accessible to everyone will necessitate breakthroughs in both Space technology and Space health, including the following essential areas:

- **Cost-effective Launch Technology:** Because the high cost of sending objects into Space is one of the most significant barriers to Space access, creating more cost- effective launch technologies is critical to making Space travel and exploration more accessible to more people.
- **Spacecraft Safety and Reliability:** Ensuring spacecraft safety and reliability is critical for

making Space accessible to everyone; research must be conducted to increase safety measures and reduce risk in space flight.
- **Improved Life Support Systems:** Another crucial area for making Space accessible is the development of improved life support systems that can sustain human life for long periods of time in severe environments, such as systems for food, water, and air recycling.
- **Telemedicine:** Telemedicine technology is critical for astronaut and future Space travelers' medical care, as it allows for remote monitoring and diagnosis of medical issues in Space.
- **Human Adaptation to Space:** Because humans are not naturally suited to the Space environment, studies on the effects of long-term exposure to microgravity, radiation, and other Space-related factors are required to better understand how humans can live and work in space.

- **Robotics and Automation:** The advancement of robotics and automation technology will lessen the need for human interaction in Space and make complicated operations in Space easier and more efficient.

To make Space accessible to everyone, we will need a persistent effort in the fields of Space technology and health, as well as collaborations involving governments, Space agencies, commercial companies, and research institutes.

Benefits of Space Democratization

The democratization of Space travel should not be viewed as a burden on Space health, but rather as a solution: the amount of data and knowledge gained will help us make Space safer for humans, reach further destinations faster, while making significant improvements for humans on Earth.

This is an excellent example of how research conducted in Space has assisted us in managing and solving numerous health issues on Earth. Similarly, we may soon discover how the effects of microgravity can transform disabilities into abilities in Space, which could significantly improve the quality of life for certain groups of people.

The possibilities are boundless, and we have much to learn; democratizing Space can help create a more inclusive and fair future for all by making Space technology, health, and resources more accessible.

CHAPTER 4

ASTROMEDICINE'S FIRST PILLAR: EDUCATION

Astromedicine, as a field, has the potential to revolutionize the way we approach healthcare and Space exploration; however, in order to make significant progress, certain principles must be followed. We have previously discussed and demonstrated that the five key principles essential for *Astromedicine* are collaboration, communication, innovation, interdisciplinary research, and a focus on practical applications.

Despite the fact that the scope of various teams working on Space exploration overlaps significantly, there appears to be a lack of communication and collaboration among them, which is a major roadblock in Space Medicine, but significant opportunity in the development of *Astromedicine*. The corresponding teams have classically not sat down together and had the necessary conversations to progress, unless there was a major sociopolitical reason to do so (aka Cold War). My proposed solution to this problem is to create a new field: ***Astromedicine***, which would consist of experts in all the fields of Biology, Medicine, Engineering, Business, Finance, and Astrophysics. Each of these fields has a unique perspective to bring to the table and combining them can lead to innovative and practical solutions.

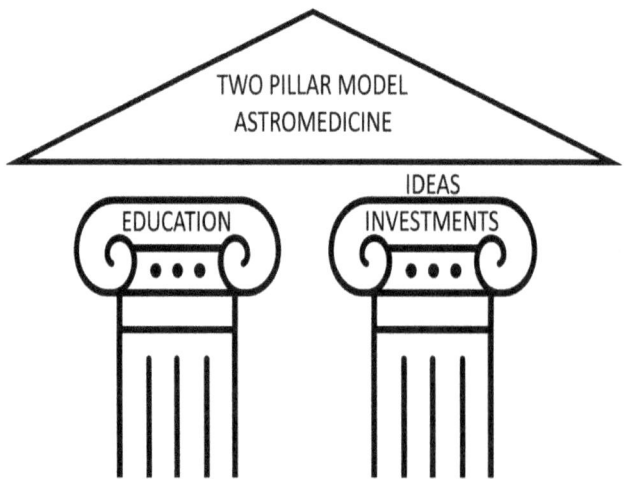

Astromedicine: redefining Space Medicine

We can identify the areas that must be focused on and allocate resources accordingly by breaking down *Astromedicine* into its various sub-disciplines, mentioned above. The five elements that comprise *Astromedicine* are critical for ensuring the safety of Space exploration.

- **Medicine**

Medicine can help understand the effects of Space on the human body and develop countermeasures for Space-related health issues. Furthermore, an Astromedical professional should understand how medicine works at the organ and human levels, including physical and mental health, as this knowledge is critical in understanding how a person will fare in space. For example, how does depression manifest in space? What happens to human blood pressure in microgravity? Can we sustain human pregnancy in space?

- **Biology**

Through the study of Biology, we will be able to understand things like; how pancreatic cells react to specific foods in space and how cardiac and muscle cells respond to microgravity over time.

Furthermore, Astromedical professionals should understand how clinical biology works at the organ and human levels, including physical and mental health, as this knowledge is critical in understanding how a person will fare in Space. For example, how does depression manifest in space? What happens to human blood pressure in microgravity? Can we sustain human pregnancy in space?

- **Engineering**

Engineering knowledge is essential for developing the necessary tools and devices for biomedical and medical applications. It is critical to have a solid understanding of mechanical and

Astromedicine: redefining Space Medicine

electrical engineering to safeguard what has been learned from Biology and Medicine and to protect human life.

For example, understanding how Space suits work and how to develop different types of Space suits for different people is required. Developing a Space suit compatible with the human body necessitates understanding Engineering concepts such as material science, thermodynamics, and fluid mechanics. Medical professionals who have this knowledge can participate in designing and developing Space suits that cater to the specific needs of different individuals.

Another critical aspect that Astromedical professionals must consider is how to make space travel more inclusive and accessible to all individuals, regardless of their physical and mental health. To do so, they must develop capsules, rockets, and settlements that are friendly to the population's mental health, which medical professionals can design and develop in collaboration with engineers.

- **Astrophysics**

The next critical component in space exploration is Astrophysics, which helps us understand the laws and conditions of Space. Understanding the different environments, we are heading towards is critical, as each environment presents unique challenges. For example, a specific environment may be more susceptible to radiation that the International Space Station (ISS) has not experienced before.

To close this knowledge gap, we must conduct extensive research on Space conditions, such as radiation levels, gravity, and temperature variations, in order to design spacecraft and settlements that can withstand and adapt to these conditions. For example, if we plan to establish a Space settlement, we must be prepared to deal with unique radiation levels that do not match those on the Earth's surface.

Astromedicine: redefining Space Medicine

- **Business & Finance**

The fifth element in ***Astromedicine*** is business and finance, which connects everything else. Knowledge gained from Biology, Medicine, Engineering, and Astrophysics can be interpreted as something profitable through Business, and investors can come in, evaluate the ideas, and provide the necessary funding to make them a reality.

The Business and Finance component also allows small businesses and entrepreneurs to become involved in the Space exploration industry; the sector is not just for large corporations and governments; small ideas can also contribute to the advancement of Space exploration. As a result, we can transform ideas into reality through good knowledge of Business and Finance, leading to significant strides in Space exploration and commercial spaceflight.

The Astromedical professional's role is

critical in bridging the gap between the business and research aspects of the healthcare industry; they will have the knowledge and skills needed to navigate the complexities of the healthcare market while also having a deep understanding of clinical research and the most up-to- date medical techniques.

They will be well-versed in bringing innovative medical products to market, from performing market research to establishing strategies that appeal to investors, and their expertise allows them to develop clinically suitable solutions that can be efficiently applied in both Earth and Space markets.

Furthermore, the Astromedical professional deeply understands the unique challenges of healthcare in Space, and they will be equipped to make informed decisions about what is right and wrong for the health and safety of humans in Space, thanks to their education and training. With a workforce of these professionals, we can ensure that medical technologies for

Astromedicine: redefining Space Medicine

Space exploration are developed responsibly and safely. Additionally, by "speaking" all the previous mentioned languages (aka from Biology to Business and everything in between), Astromedical professionals will be the "glue" that holds all aerospace fields together, as these individuals will be able to communicate with everyone.

The current workforce consists of various paths, experiences, and education systems. It is critical to analyze this concept and understand how it affects the Space sector. However, with **Astromedicine**, we aim to bring in a new set of skills that will create a unique knowledge base for individuals in this field.

When someone will say that they are studying **Astromedicine**, it will mean something specific and unique to the listener. This field will not only provide students with a diverse educational background, but also a clear direction for their future career path. Graduates of **Astromedicine** will have a specialized place in the workforce, utilizing their skills and knowledge.

Finally, education is critical in shaping the workforce, and *Astromedicine* provides a unique perspective on this concept. With this new discipline, we can create a workforce that is not only diverse but also has a clear direction and purpose, and the global implementation of this educational system will pave the way for a safer and more secure future in Space exploration.

Astromedicine is an intriguing concept with the potential to revolutionize the medical industry; however, we must first educate the next generation of scientists, doctors, and researchers. Astromedical education will cover all age groups, from Pre-K to postgraduate studies, and everything in between.

We can bridge the gap between diverse groups of people who have varied pathways and educations by including *Astromedicine* into the school curriculum, which will enable them comprehend each other's perspectives and communicate successfully, even if their scopes overlap.

The focus of early education should be on introducing basic concepts of *Astromedicine*, such as Space physiology, life support systems, and Space environmental conditions. In middle school and high school, students can delve deeper into the subject and learn about practical applications of *Astromedicine*, such as how it is used to keep astronauts healthy in space, to include performance of laboratory experiments to test out theories.

Students can concentrate in *Astromedicine* and specialize in fields such as Space Medicine, Space Pharmacology, and Space Psychology at the undergraduate level, as well as participate in research projects connected to *Astromedicine* and receive hands- on experience in the field, either by working in the industry as interns or by sending actual experiments in Space.

Graduate and postgraduate studies will provide individuals interested in employment in ***Astromedicine*** with advanced courses and research possibilities in areas such as Space Physiology, Telemedicine, and Space based healthcare technology.

Astromedicine graduates will work in a variety of industries, including Space agencies, private Space companies, and healthcare organizations, as well as research and development of new technologies and treatments that can benefit people on Earth and in Space.

To summarize, the educational system of ***Astromedicine*** is critical for preparing the workforce required for the future of Medicine. By providing a comprehensive educational program that covers all age groups and academic levels, we can ensure that the next generation of professionals will be well-equipped to handle the challenges and opportunities of ***Astromedicine***.

Finally, the Astromedical professional can drive innovation in the healthcare industry and

Astromedicine: redefining Space Medicine

pave the way for a healthier and more sustainable future in Space for all humans by combining their business acumen with their clinical expertise.

V. Farsadaki

CHAPTER 5

ASTROMEDICINE'S SECOND PILLAR: THE WHEEL

As humanity expands into Space, we face numerous challenges that require innovative solutions, ranging from the physiological effects of long-term spaceflight to the logistical challenges of establishing a sustainable non-terrestrial presence. To address these issues, I have proposed *Astromedicine* as a viable and long-term solution.

Astromedicine, the application of medical knowledge and technology to Space exploration and colonization, provides a comprehensive approach to addressing the unique health needs of all Space travelers. In the previous chapter, we discussed the essential curriculum required to educate the workforce capable of making *Astromedicine* a reality. In this chapter, we will delve into the topic of investment and investigate

Astromedicine: redefining Space Medicine

how it can act as a catalyst for the sustainable development of the field.

Investment in *Astromedicine* is critical to its success; it is the wheel that propels innovation and drives progress. There are numerous areas where investment can make a significant impact, such as research and development, technology, infrastructure, and training. By investing in these areas, we can foster the growth of *Astromedicine* and ensure that it becomes a sustainable field that benefits humanity for years to come.

In this chapter, we will look at different educational and investment ideas and examine their potential impact on Astro Medicine; with the appropriate educational curriculum and investment strategy, we can take Astro Medicine to new heights and pave the road for a brighter future for humanity in space.

The first concept here is related to investments, as it has been projected that the space sector will be worth approximately four to five trillion dollars by 2030. However, despite the immense potential for growth, Space health issues have not been given enough attention by investors and this is our first challenge.

The second challenge is the disconnect of Space technology and Engineering from Space health and Medicine, which is a significant issue that must be addressed. While the focus has been primarily on developing new technologies for Space exploration, little attention has been paid to Space health and medical issues. As a result, it is critical to integrate Space health and Medicine into the sector's growth plans and investment strategies to ensure sustainability.

To accomplish this, we must encourage more investments in Space health and Medicine research, development, and innovation; the Space industry must collaborate with experts from various fields, such as Astromedical professionals, engineers, and scientists, to develop a holistic approach to Space exploration that prioritizes human health and safety; and it is critical to educate and raise investor awareness about the importance of Space health.

Furthermore, the major challenge in the emerging Space health market is that potential investors are still waiting to see where they can make a difference; they are carefully monitoring the market, looking for the right opportunity to invest; and it is at that point, when they finally decide to invest, that the game-changing idea of **Astromedicine** will be already set in place, waiting for them.

The other challenge is the sheer number of new ideas emerging every day in the Space health and pharmaceuticals sector, ranging from medical technologies to pharmaceuticals in Space. While many of these innovative ideas appear promising, they frequently lack the necessary funding to bring them to fruition, resulting in an increase in startups, each with a single brilliant idea but struggling to secure the necessary funds to develop it further.

Startups are having difficulty breaking through and entering the market because they have no idea how to reach out to and secure funding from investors, which is exacerbated by the fact that many startups require significant funding to develop their idea, create a proof of concept, and develop a prototype.

As a result, we face several critical issues in this sector. First, investors require guidance on how to enter the emerging space health market. Second, innovative startups require funding to develop their ideas and become viable businesses. By addressing these challenges, we can unlock the potential of the space health market and usher in a new era of innovative medical technologies and pharmaceuticals.

What is the relationship between these pillars?

Astromedicine will aim to establish a consortium where startups and investors can easily access each other, thereby creating a spinning wheel of ideas and investments.

Astromedicine: redefining Space Medicine

Astromedicine will expect that through bringing together investors and entrepreneurs, it will not only keep the wheel moving but will also give back to investors as firms grow, allowing the *Astromedicine* model to be sustained in the long run.

In addition to making space travel safer, the *Astromedicine* model intends to inspire fresh concepts that have never been used on Earth, which, when combined with investments, will have a significant impact not just on Space, but also on Earth.

The success of the Astromedical model will benefit not just the Space industry but also society, since it will inspire advancement and promote economic growth in a variety of industries by encouraging collaboration and innovation.

The Space health and Medicine industry has gained traction in recent years, with an increasing number of investors and companies showing interest in this field. According to Allied

Market Research, the global Space health market is expected to reach a whopping $2.8 billion by 2026, growing at an 8.2% CAGR from 2019 to 2026.

The term *"Emerging Space economy"* refers to the growing economic activities related to space exploration and utilization, such as the development of satellite technologies, Space tourism, and the commercialization of Space resources such as asteroid mining. The Space economy has been expanding rapidly in recent years, driven by increasing private company involvement and the decreasing cost of space exploration and utilization.

When it comes to expertise, it is critical to identify the type of skills and knowledge required to fill key positions; in this case, we need a workforce that is educated and trained to work in the field of *Astromedicine*; education is the first pillar of our model.

Astromedical professionals will be responsible for evaluating ideas and determining

Astromedicine: redefining Space Medicine

whether they have the potential to be useful in Space, including settlements, colonization, habitats, lunar clinics, smart clinics, insurances, and space Bioethics.

This workforce must be able to assess the viability of these ideas, examine the risks, and make educated decisions about whether or not to invest; they must also be able to recognize the most promising ideas, comprehend market potential, and devise strategies to bring these ideas to fruition.

With their expertise in Biology, Medicine, Engineering, Astrophysics, Business, and Finance, these graduates will be instrumental in driving innovation and growth in the Space health market. C-suite leaders who possess a deep understanding of the workings of the market and the science behind it will be in high demand as companies seek to innovate.

The Chief Medical Officer (CMO) and Chief Scientific Officer (CSO) are two such key positions that require specialized knowledge and skills. These executives will be responsible for leading startup consortiums and securing funding for their ambitious ventures, as well as guiding the development of new products and technologies and ensuring their commercial success.

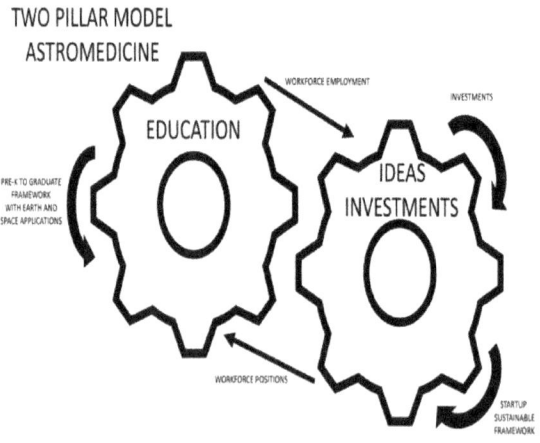

Small companies with innovative ideas are poised to make a big impact in today's rapidly evolving landscape, and having a team of experts with diverse skill sets is critical for success. A C-suite team that includes someone with a deep understanding of both the market and science will be a valuable asset in achieving profitability and ROI.

The proposed model for developing this workforce should be fully operational within the next five to ten years, coinciding with a period when private and governmental Space missions are expected to increase significantly. Companies will embrace this model and hire executives with a thorough understanding of the Space health market to stay ahead of the curve.

In conclusion, the potential for growth and innovation in the Space health market is enormous, but it necessitates a new breed of executives with a unique combination of skills and expertise. The proposed model offers a solution to the challenge of finding qualified candidates for these positions, and companies can achieve success with the right team in place.

CONCLUSION

As we conclude, it is crucial that we think back on our trip together as we approach the book's end. We have discussed a wide range of subjects, including the development of *Astromedicine*, a thorough examination of the past, the democratization of Space flight, and ways to fill in the knowledge and technological gaps. I wish to summarize the key ideas covered throughout the book in this section and leave you with a personal message of inspiration and hope for the future of *Astromedicine*.

The early chapters covered the development of Space medicine and described the scientific and technical advances that have influenced how we view Spaceflight. We saw where things went wrong and emphasized the medical community's somewhat sluggish adoption of these innovations, which has left us juggling the difficult problems of human health

Astromedicine: redefining Space Medicine

in Space. We emphasized the need for cross-disciplinary cooperation to establish a safer environment for Space travel through personal experiences and professional views.

We then discussed the idea of *Astromedicine*, a concept and term I created to draw attention to the distinct medical problems caused by how the human body interacts with Space. This sparked a conversation about the gaps in our present knowledge and the necessity for a new workforce of professionals that will have a set of skills in disciplines including Biology, Medicine, Engineering, Astrophysics, Business, and Finance to collaborate.

The argument for the necessity for *Astromedicine* establishment was made using the idea of democratization as its motivating factor. It is critical that we address the plethora of potential health risks as we advance toward a time when everyone can fly to Space. We can create comprehensive solutions that make spaceflight safer for everyone by combining the study of

Biology, Medicine, Engineering, Astrophysics, Business, and Finance.

The two pillars of my model investments ideas wheel and education are at the heart of this book. It is crucial that we enlist investors who understand the potential of this industry to advance and sustain **Astromedicine** into the future. By doing this, we may support the study and creation of ground-breaking discoveries and breakthroughs. On the other hand, education is crucial in developing the upcoming generation of **Astromedicine** specialists that will support the investments-ideas wheel. We can guarantee a future workforce that is prepared to handle the challenges of spaceflight by ensuring that the new generation has the knowledge and skills that will be required.

I have made an effort to explain in these pages both the significance of **Astromedicine** and the pressing need to address the current Space health and "Space Medicine" problems. We must make sure we are ready for the medical

Astromedicine: redefining Space Medicine

challenges that come with spaceflight as we approach a new period in human history where Space travel is becoming more widely available. Future generations can be prepared to live in a world where spaceflight is not only feasible but also safe, by making investments in research, promoting collaboration, and raising awareness.

I'll leave you with a special note of encouragement before I go. Although *Astromedicine* is still in its infancy, there is tremendous room for expansion. I urge you to keep an open mind, be curious, and be excited about the possibilities that lay ahead. Together, we can influence the course of Space exploration and make sure that everyone who travels into the vast unknown does so with the knowledge and tools essential to sustain their health and well-being.

Keep in mind that the sky is not the limit, but merely the beginning.

Welcome to Astromedicine!

V. Farsadaki

AUTHOR INFORMATION

Dr. Vanessa Farsadaki is the preeminent thought- leader towards advancing the discipline of Space Medicine. A proud American citizen of Greek descent and a Medical Doctor, Dr. V as she is known colloquially, is on-track to become Greece's first-ever Astronaut in history. She is a scuba diver, a pilot in training and a skydiver. As the President and Managing Partner of Space Exploration Strategies LLC, Dr. V's impressive bona fides include advanced degrees in Biology, Genetics, Astronomy, Astrophysics, and Business Leadership. She has authored and co-authored a plethora of articles advocating for advancements in her field of expertise, and her deep experiences with Radiation Exposure and Protection have made her a sought-after advisor on high-end programs of note for the gravitas she brings to the discussions.

Additionally, her fluency in 18 languages is a testament to her ability to engage with a

Astromedicine: redefining Space Medicine

broad range of audiences, and thus she continues to serve as a keynote speaker at numerous fora to include the prestigious Kings College in London, England and many other universities internationally. She has won numerous awards among which the International Trailblazer award for her futurist work in Space Medicine. She is a British Interplanetary Society fellow and the youngest Space Ambassador to be ever recognized by the National Space Society.

V. Farsadaki

www.ingramcontent.com/pod-product-compliance
Lightning Source LLC
Chambersburg PA
CBHW071059240526
45471CB00016B/2156